"*In Olympus, every famous spring had its own naiad (nymph). If a human being entered the spring, he would either come out cured – if the naiad was merciful – or he wouldn't come out at all if he did something the nymph deemed offensive. The naiads were beautiful, alluring creatures. Wonder why? Well, they were always around water.*"

**Almost Everything
about
Water Therapy**

Author: Constantin D. Cerbu
Translator: Armina Sîrbu
Design and Publishing: Daniel Sîrbu

Copyright and Legal Notice

This publication is protected by international copyright law, federal, state and local laws, and all rights are reserved, including resale rights. No part of this document may be reproduced, distributed, resold or transmitted in any form or by any means, including photocopying, recording, or other electronic or mechanical methods, without the prior written permission of the author.

Any trademarks, service marks, product names or named features are assumed to be the property of their respective owners, and are used only for reference. There is no implied endorsement if we use one of these terms.

Copyright © 2013 Constantin D. Cerbu
All rights reserved worldwide.

ISBN #: 978-1-300-61082-3

Table of Contents

Introduction	7
The amount of water in the ground	25
Springs, lakes and mud	26
Drinkable water	34
Be informed!	42
Water from refreshment drinks	43
Processed water therapy	46
Healers and water	51
Magnetized water	52
Solarized water	58
Living water therapy	59
Distilled water	63
Pi water	64
Kangen water	65
Silver water	67
Water enriched with coral calcium: CORAL-MINE	73
Shungite water	73
Treatment with clay that has been dissolved in water	74
Water and deuterium	78
Water low in deuterium	80
Hunza – a miracle	82
Regular water	86
Bibliography	94

Introduction

Earth is the most blessed planet of our Galaxy. Not only does it have an atmosphere and just enough sun exposure, but it also has large volumes of water. None of the known planets to date have as much water as there is on Earth. Actually, it was water that shaped the planet and, from certain perspectives, it is responsible for the processes that shaped it as it is today. But where does it all come from? Water was necessary. Billions of cubic meters of it (1234 billions) and maybe more, who knows the numbers? Some say that the first drop of water on Earth came from the telluric mantle, for it had a reserve of approximately 20×10^{18} tons. To form the hydrosphere, $3,4 \times 10^{18}$ tons "left" initially and approximately 10^{10} got "lost" in space through dissociation and our hydrosphere was left only with $2,4 \times 10^{18}$ tons. "As the planet changed, the conditions allowed water to be found in every layer of the planet, from the telluric mantle up until the upper atmosphere. This mobility turned water into the main link between the various subsystems of the planet for energy and matter exchange. Therefore we could say that the hydrosphere is the planet's circulatory system. It may be possible that water played a key role in the forming and organizing of the planet's geography even since Earth was beginning to take shape. The surface of the Earth's crust is constantly in contact with water. Because of its particular landscape, meaning the great depressions of the oceans and the seas, and due to the weather conditions, the hydrosphere has three distinct subsystems: oceanic, continental and glacial. From the small lake that lays hidden between tall mountains to the raging torrent that comes crashing down the rocks, from the colossal water "blankets" of the Amazon to the icy Antarctic "continent" with its floating pack ice, they are all linked to the Planetary Ocean, highlighting the continuity of the hydrosphere."

Earth had little water in its early days. The water was brought later on by comets and meteorites, which carry water in an icy state. We know today that water represents only 0,6% of the planet. The 2003 Deep Impact mission successfully managed to hit a comet with a missile, thus confirming the presence of water in comets, although it seems somewhat different from Earth water. In the year 2000, the traces left by a meteorite in the sky have been photographed. Its remains were studied and it was found that the meteorite was made 20% of water. This would imply that, in the past, there could have been a massive meteor shower on Earth. It would seem that, at a certain point, Jupiter changed their trajectories and they came across our planet. The building of water lasted hundreds of millions of years. Almost 150 million, as it was found when analyzing zirconium silicate. And yet, there's still more water on the asteroids near Jupiter. Should we wait for them? No, we have enough. We know that water built up at the surface and in the ground. There has been a recent discovery of an aquatic environment isolated from the rest of the world for 15 million years. This environment is Lake Vostok, located in Antarctica, below a 4 km sheet of ice. This lake is one of the largest bodies of water on Earth, measuring over 10.000 square kilometers and has a depth of 800 m.

One impressive thing about ancient Greek mythology is that it has a large number of deities associated to water in all of its forms and states. Therefore, along with the popular Poseidon, god of the seas and oceans, there are lesser known gods like Okeanos and his sister Tethys. Then there's Glaucos, the god of marine life, Typhon, the hurricane god, Eol, the god of wind etc. Now that we think about it, it isn't that impressive at all, knowing the colossal part played by water in nature: developing and supporting life on Earth.

Remember the wonders of the water cycle from school? Water rains down on the ground, where it's picked up by rivers and carried to lakes and seas and oceans, from where it evaporates right back into the atmosphere only to fall down again as rain. Actually, this whole cycle is a metaphor for the circle of life. When you enjoy a glass of water you could say Caesar or whoever you think of drank a tiny part of it once.

The wonder here is that water resists gravity when it evaporates. Has anyone found out why? It's hard to say. The full answer isn't that this happens because water vapors are less dense than air.

In the words of Leonardo da Vinci, "water is the driving force of all nature". This is true, given that life was born in water. By extension, every living being is the work of both Sun and water. Water supports all life on Earth and it also is one of the most perfect solvents nature has created, for it doesn't alter the biochemical properties of countless substances that dissolve into it. Chemically speaking, water is a mineral compound consisting of hydrogen and oxygen (H_2O) which is found in three states: solid (ice), liquid and gas (steam, water vapors). When at room temperature, water is a liquid with no color, smell or taste. It's the most widespread compound, covering about 75% of the Earth. Life depends on water in all stages of its existence; this ability that water has to dissolve countless substances is, perhaps, its most important feat. Life is believed to have emerged in the Planetary Ocean, and all living organisms use wet environments to conduct their biological processes. Water molecules are asymmetric and, therefore, electric dipoles. Because of that, the hydrogen links between molecules of water (solid or liquid) are vital pieces that hold them together. It's because of these links that water has such a vast and complex number of properties. When in the presence of salts and other soluble substances, water goes through a process of dissociation into H+ (or H3O+) ions and OH- ions. Therefore it can act both as a base and an acid. Water is a part of chemical links in many salts and minerals, known as water of hydration. Water covers more than 2/3 of the planet and almost 2/3 of the human body. The water in the human body is, as in nature, in a permanent cycle. Each day we lose about 500ml of water through perspiration, 350ml in the air we exhale and 1,5l in urine. But it doesn't just go from inside to outside, going in the opposite direction as well.

The human body's sources of water are: liquid water (ingested through the mouth) and water synthesized from food through oxidation. Food is an important source for water, for it is widely known that some fruits and vegetables have water in a percentage up to 95%. Another interesting fact is that the human body can, within certain limits, produce the water it needs. This product is known as endogenous water and it is formed during the metabolic processes through polymerization. In such a process, 100g of fat produce 107,1 ml of water, 100g of protein produce 41,3 ml of water and 100g of alcohol produce 117,4 ml of water.

The process of involution, known as the aging process, is tied directly to the body being less and less capable of processing and retaining water. A 3-day-old embryo contains 97%, which drops to 81% at 8 months. The rate of water loss is constantly growing faster, making that by age 30 the percentage is 60% and by age 70 it is about 46%. As we get older we constantly "dry up"; like trees.

Therefore, water is the reason that the smooth skins of our youth roughen up when approaching old age.

Without getting into the more technical details, we can state that water behaves differently depending on its source and external conditions. Differences can be observed on:

- the speed it evaporates with,

- the way it produces different polarmetric deviations when immerged in sugary solutions of the same concentration,

- the different responses water produces when exposed to identical magnetic fields,

- the way it behaves in electrolysis,

- the different ways water behaves when different gases are inserted into it (O2, H2, CO2 etc.),

- the different degrees of solubility;

- the behavior it manifests when hit with a laser beam,

- the angle of deviation corresponding to polarized light that differs when faced with recipients water of the same temperature,

- the different results obtained on a spectroscopy using a laser that emits pulses once every picoseconds.

Nature has provided all living things with techniques of all kinds to face a lack of water, from the most simple to the most complex. Observing human behavior, we begin to see several interesting aspects about the water consumption. For example, in Sahara, under excessive heat conditions, the human body can lose up to 12 liters of water per day, meaning 1 liter of water each hour. Unlike animals, that drink only when becoming thirsty, people can drink out of habit (more like "bad habit", as this doesn't come without a cost). At the same time, people living in dry, barren lands have adapted to water shortages, consuming only about 1 liter per day. Even so, thirst is a major problem of human biology, for there is no worse feeling than thirst.

Water has three dominant characteristics: source of life and healthiness, agent of physical purification and regeneration and curative, therapeutic factor. A part of nature, liquid and at the same time a perfect solvent since the biochemical properties of the substances it dissolves remain unaltered, water symbolizes division. At the same time, it also symbolizes cohesion, since water is homogenous as a whole. On the other hand, water is considered to be a primal, fundamental element, the original building block. This definition of water as a primordial element, as a point of origin, is almost universal. Old Hindu manuscripts note that "everything is water".

Babylonian astrology claims that at the beginning of all things, when there was no sky and no earth, everything existed in an environment that was unvarying, the primordial waters. Before all things, though, water is food. Vital food. That being said, we'll continue on making some practical recommendations concerning an everyday action: drinking water. It's an action that our physiological needs had us practice so often that we've developed certain automatisms. However, this action can too be improved by creating certain premises concerning integrating water into the body in a way that it keeps its most natural parameters. It's important that we drink just the right amount, no more no less, and that we apply some of nature's techniques to energize the water we consume.

If you keep water in a bottle, shake up the bottle a little, so as to imitate the rocks it crashes down upon in a waterfall. Traditional culture says that water that rhythmically falls on the rocks from the height of a waterfall has revitalizing properties that can't be found in water that runs smoothly.

Before drinking it, let the water bathe in the sunlight for a few minutes. "Set a pattern for drinking water during the day: drink just before a meal, drink as little as possible while eating and don't drink at all in between courses" are some of the recommendations issued by Kneipp.

Water is also a cure, the primal cure. We're speaking here about mineral water used either internally or externally. However, normal water, water that comes from springs, rain or atmospheric condensation is just as important, therapeutically speaking. A very popular treatment in the past was the hydrotherapeutic treatment of abbot

Kneipp. It consisted of walking bare-footed through grass covered in morning dew. This way water triggers essential modifications in the body, not only affecting the skin, but soaking through to many tissues and systems. At the time, Kneipp was known as a "roi des eaux" by the French, as a "Wasserkönig" by the Germans, these names meaning "water king". In a social psychosis, people branded him as the best doctor of the time, although he wasn't really a doctor.

Countless doctors travelled to Worishofen, the German city in which Kneipp worked, to witness first-hand this new method of treatment.

Cold water treatment, a process with significant galvanizing properties, was also accompanied by a strict diet and outdoor exercise. Kneipp's method doesn't qualify as a cure-all remedy, of course, but it does qualify as a valuable therapeutic method, primarily due to the vasodilatory and vasoconstrictive effects dew and cold water have on the body and the after-effects that derive from them. Secondly, it's due to the way cold water stimulates the countless receptors and nerves found in the derma, thus helping the stimuli reach the central nervous system.

On the other hand, Wilson made some experiments

between 1906 and 1908 which determined that the atmosphere has an existing electric potential that has a variation of minimum 100 volts/meter. The electric potential of a point located 1 or 2 meters about the ground can suddenly reach between 3000 and 5000 volts (that's ten times the normal value) and then suddenly switch between negative and positive values of the same scale. The maximum value is reached at 7:00 PM London time, no matter where on Earth we decide to do the measuring. The minimum is at 4 AM.

Therefore we can come to the conclusion that a person sitting in a field, with its feet in a wet, conductive environment, will be passed by discharge electric currents. These currents work just like electrotherapy, a thing Kneipp couldn't have known about, but a thing science was going to use later on to cure many conditions.

About the same conditions are cured when following a water therapy plan: saltwater baths, seawater baths, mud baths, etc.

In Tibet, the hermits are known to sit on the bare ground, lightly clothed, subjecting their bodies to the passing electrical currents that, although their voltage is low, produce positive health effects after prolonged exposure. On a related matter, walking barefoot was a rite of passage in ancient Greece and, according to the Greek historian Pausanias, in Sparta and Athens students were walking around barefooted all year long, earning the right to wear footwear only when they were 18 years old.

Plutarch, another famous historian, writes about how even Socrates himself walked barefoot, convinced that it helped him keep a sharp mind. Seneca believed the same.

At its turn, Horatio wrote that his healer made him walk barefoot, a treatment that helped him recover from serious condition.

As far as we know, many balneal resorts around the world treat people based on these very principles.

But let's return to the symbolism of water. In many cultures and mythologies water is regarded as a symbol of health, strength and joy.

In many communities water is a symbol of hospitality, resulting in the ancient custom to offer guests a glass of fresh water when welcoming them into your home.

Because of its many purifying qualities, water helps heal and rejuvenate.

No wonder that the first medical encampments first appeared on the waterfront. Water immersion is praised in the culture of many peoples for its rejuvenating properties and it is believed to make people healthier and live longer, making it a part of countless rituals.

Moving along to hydrotherapy, we'll begin by giving the medical definition of the term: a branch of medicine that relies on the external use of water to treat or prevent disorders, illness or injury. Whether the patient is exposed to water by bathing, showering, applying gauzes, hot towels or other spa-specific procedures, it stimulates the body in the same ways, the two most common being thermal excitation and mechanical excitation. Skin acts as an intermediary between water and the body, being the organ that directs these stimuli to best benefit the body. The level of thermal excitation corresponding to the water is conditioned by its temperature. Therefore, depending on the temperature of the water, there are three branches to hydrotherapy:

- cryotherapy, when the treatment is done using ice;

- actual hydrotherapy, treatment by using water of a temperature that stays below 42°C;

- thermotherapy, which uses steam at a temperature between 42°C and 80°C or delivers the heat using overheated air, which is the case of traditional steam baths or saunas.

At its turn, mechanical excitation has its own way

of stimulating the skin and, by extension, the body. This excitation is linked to the mechanical pressure of the water, which increases when an important part of the body is submerged. Hydrostatic pressure depends on water density and the height of the water column on top of the body. Baths also have an influence in the dynamics of blood flow, relaxing the veins and increasing the amount of blood going to the heart. This is the reason the heart becomes weary while bathing. Short baths, alternatively hot and cold, lead to an increase in skeletal muscle tone, as well as an increase in the overall muscle capacity.

Regarding the effects baths have on metabolism, cold baths have an invigorating effect, affecting both assimilation and dissimilation, thus intensifying the metabolism. Hot baths stimulate an increase in metabolizing fat, as well as carbohydrates, and if the bath is prolonged a protein dissimilation occurs, resulting in an increased subsequent elimination of uric acid and urea. Normal baths don't produce any effect on the metabolism. Typical warm baths inhibit the nervous system, which explains the feeling of calm and sleepiness warm baths give.

The most accessible ways for water therapy are steam baths and saunas, the two forms of hyperthermal water therapy. The human body is a vast system of canals constantly teaming with liquid and gas. We're going to talk next about the capillaries, the vast network through which 80% of the total blood travels.

Those among us who have been in a sauna or a steam bath before may have seen that their entire skin turns crimson-red. That's because of the capillaries in the skin are dilating. They may have also seen that, when diving into the pool of cold water, the skin returns to its normal color or, quite the contrary, becomes even more white and pale. That's due to the vasoconstrictive effect that the cold has on it. This "vascular gymnastics", along with turning the skin into an excretory organ, is one of the great properties of steam baths and saunas.

At the beginning of the 20th century, Alexandre Salmanoff, a French doctor of Russian origin, has come up with a therapy method that became extremely popular at the time.

Hyperthermal capillary therapy can be easily tried at home. It consists of taking showers or baths before bed, using water between 36°C and 40°C. They must be done gradually, initially for 3 to 5 minutes and then, day by day, increasing their length to somewhere between 20 to 30 minutes, depending on how the individual tolerance.

The benefits of hyperthermal water therapy include

lower levels of fatigue, sleep induction and overall tension reduction, with all its effects on the body. These benefits are, in part, possible due to the psychological influences heat has.

Steam baths and saunas are just ways in which hyperthermal water therapy has been vulgarized.

Steam baths came into being due to the knowledge of Eastern or Northern peoples. Not only does their vast experience endow steam baths with healing the body, but it makes them also heal the spirit, removing tension and restoring balance to the mind. People noticed that these steam baths, like the sauna, their modern equivalent, had a purifying ability. Because of this, steam baths quickly became important public whereabouts in the ancient world. In Roman times, the forum, the temple and the thermae (the steam baths) were the three legs of a tripod that held in place the entire public activity of the city. Because of that, these buildings were always built next to one another, right in the center of the city. Their whole civilization was the tripod, and those were its three legs.

The average Roman left his clothes in an apodyterium, which was a locker room, and then entered a cool chamber known as the frigidarium, where he performed his first ablutions. After that, he passed into a chamber that had tepid water – the tepidarium –, where his body enjoyed the beneficial influences of the moist, warm air. After the tepidarium, he went into the caldarium, the place where he took the actual bath, the bath that will make his body perspire abundantly. The thermae were also fit with massage chambers, where the body was massaged using various scented oils, which not only imbibed the skin with a pleasant, alluring smell, but also restored its flexibility, which was heavily affected by this excessive dehydration. Increased sweat secretion in steam baths and saunas helps eliminate large amounts of metabolic waste products. The thermae also had chambers for gymnastics, rest and relaxation. By the year 33 B.C., in the time of Agrippa, Rome had 170 private thermae, in addition to the main, central ones.

The modern sauna took the concept of the Roman thermae and, by amplifying and prolonging the effects, improved it. When exposed to steam and sweat, the skin isn't only a protective organ. It becomes an excretory organ and a breathing organ as well.

Hydrotherapy causes the skin to produce complex and profound modifications within the body, mainly because of the heat of the steam or that of the overheated air, as well as the low temperature in the cold water basin one dives into when getting out of the steam room. The body's fine-tuned thermoregulatory systems will immediately trigger vasodilation or vasoconstriction.

One can note that the active dynamics of these opposite processes does indeed resemble "vascular gymnastics".

However, there are some prerequisites for those wishing to go to a steam bath or to a sauna:
- have a general good health;
- have a certain adaptability of the body;
- have the ability to follow schedules and plans.

If these conditions are met, the steps that follow are:
- the first step is getting a medical consult;
- begin the steam bath in short sessions of 3 to 5 minutes;
- gradually increase the time period;
- once out of the steam room, get into a shower and adjust the temperature to normal (or even cold), then dive into the pool and swim around for a bit;
- when done, take a break; rest comfortably for at least 10 minutes.

It's not recommended to partake in this process on a full stomach or under the influence of alcohol.

The amount of water in the ground

The amount of water in the ground is given by the ratio between the amount of water that enters the ground and the amount of water that leaves the ground through evaporation and groundwater drainage. It reveals the ground's degree of humidity and the cycle of groundwater.

Aside from the minerals and organisms that reside within the ground, there is also water. Once it reaches inside soil, water is subjected to certain forces: gravity, the capillary force, absorption etc. As a result, water is retained within it. Therefore, after a rough assessment of the degree of humidity, we notice the following types of soil:

- dry (releases dust, darkens when wet)
- moist (it doesn't release dust and doesn't dampen the filter paper when pressed against it, but it does leave a cold feeling in the hands when squeezed);
- damp (the soil sample moistens the filter paper when pressed against it; it lightens in color when dry)
- humid/ very moist (the handful of soil leaves a trace of water that glitters on the palms and the sample dampens the filter paper);
- wet (water drips from the handheld sample).

Springs, lakes and mud

Some traditional ways of therapy rely on using springs, lakes and mud for their therapeutic purposes. Mineral waters, lakes and curative muds were always considered to among the mysterious children of mother nature.

Hot springs, for example, aren't held to such high esteem just because they have therapeutic properties, but also because they have two defining psychological traits: their purity on one hand and their origin is the womb of the Earth on the other. Once they reach the surface, these waters "intensely manifest evolution or aging processes", as noted by Dubot and Fontan in a technical study.

As water has been known for millennia to be the main element in laundering, it wasn't long before thermal or mineral waters gained the same reputation, only this time the cleaning was taking place inside the body.

That's how places like Vichy (France) or Olănești (Romania) gained a reputation for having waters that "wash clean" the liver, the same going for places like Evian or Căciulata (Romania), which had waters known to "wash clean" the kidneys. Doctors agree that these therapeutic effects aren't due just to the mineral elements contained in these waters.

Therefore, it has been found that mineral waters, no matter how different in their mineral count, can have the same therapeutic effects.

Bachelard believes that, aside from dissolving, eliminating, releasing and cleaning the toxins from organs and tissues, they fill us with a sense of primal tranquility. On the same note, Guy Ebrard declared that "In these days, balneology is a way of returning to nature".

Romanian balneotherapy was always widely known and appreciated. History says that Napoleon the 3rd used water brought from Căciulata to treat his kidney stones and that pope Innocent VII spoke about the miracle waters of Felix in a letter of his.

Man has always loved springs, sometimes merging this admiration with the sacred, according to an inscription in Băile Herculane.

Sanctifying springs is common to many cultures, as many saw them as sources of the clearest, purest water. Many cultures claim spring water is, like rain, flowing celestial blood. Sometimes, certain taboos were born around springs. The Mayans of Central America forbade fishing in the holy water near springs and cutting down trees that cast their shade over them. Spring water was always considered to be the most pristine water in existence, the very substance of purity itself. In some mythologies, it is believed that spring water, when drunk directly from the spring, grants passage to the kingdom of heroes. Springs are also a symbol for the origins of life, power, grace, intellect, happiness and many more.

While springs and their water are surrounded by many old myths, there are even more surrounding mineral springs, springs that truly live up to their name of 'source of healthiness'. The idea that mineral water only releases its curative properties when drunk directly at the spring was passed on for generations. Each mineral water has its own personality, its own lasting individuality, both physically and chemically speaking. That being said, each one contains minerals in different amounts. However, that doesn't mean that the dominant mineral surpasses the effects of the others, as they all affect the body in a specific way. Furthermore, these secondary elements can change the effects of the dominant mineral, both synergistically and antagonistically. Thus, when we drink a more sulphurous mineral water, we shouldn't think that we will benefit only from the effects of sulphur, but from the effects of the secondary minerals as well. Then, the elements of mineral water are found in a stable relationship to one another, both physically and chemically, keeping a certain ionic equilibrium in their place of origin. When water is removed from this place and transported elsewhere, the aforementioned equilibrium gradually disappears as these ions become inactive, whether we notice it or not.

Recent tests have shown a separation of certain residues found within the bottles. Furthermore, some salts that were known to dissolve completely in the water at the spring now are unable to do so even when shaking the bottle.

At the spring, iron is dissociated, under the form of divalent ions. It will not precipitate, keeping its curative properties. When mineral water comes into contact with a different environment (in this case the atmosphere), it comes into contact with other temperature, pressure and even magnetic conditions, factors which can change the polymerized state of the isotopes and the colloidal state of the composing elements, therefore changing the therapeutic action. Depending on their purpose, balneal therapy can be divided into three categories:

- prophylactic
- curative
- recovery

Mineral water comes from natural mineral springs or from human drilling. For water to be internationally defined as 'mineral water', it must fulfill one or more of the following conditions:

- have a ratio of at least 1g of typical salts to 1 liter of water;
- contain an accepted minimum amount of chemicals that have a discernible pharmacological action;
- incorporate gases like carbon dioxide, hydrogen sulfide and radon in acceptable levels of concentration;
- have a scientifically proven curative action;
- have a temperature greater than 20°C at the spring.

It's good to know that some natural springs are lightly mineralized (oligomineral water) that have curative properties and are included among healing mineral waters, although having a low saline content.

Let's move on to the origins of mineral water. They get their chemicals from the underground rocks they've come in contact with. Classifying them by water source, there are three types of mineral springs:

a) Juvenile springs bring to the surface water originating from condensation within underground volcanoes. The water comes from a great depth and its mineral content is low. Juvenile waters flow through the regions with recent volcanic activity.

b) Fissure springs bring up rain water that permeated

the ground through permeable rocks or by moving through the rifts in impermeable rocks. Because it's always in contact with the rocks while flowing underground, the water dissolves some chemicals off them and when it emerges back to the surface, it can be classified as mineral water. This high solubility rate is given by the low quantities of CO_2 rain water picks up from the atmosphere or from the surface of the ground. The water from fissure springs has various levels of mineralization, depending on the types of rock water seeped through. Water temperature also varies, depending on the depth it comes from; when it comes from a greater depth, it comes as thermal water.

c) Mixed springs, in which juvenile water mixes with rain water deep underground.

d) Artesian springs occur when the groundwater reserve is surrounded by a confining layer.

Such confined groundwater reserves are sometimes in the vicinity of oil or natural gas reserves. Artesian waters sometimes contain salt with large concentrations of either iodine or bromine.

Mineral water can be classified by temperature, osmotic concentration and chemical content.

According to its temperature, mineral water can be:
- cold, with a temperature below 20°C;

- hot/ thermal, with a temperature over 20°C. The limit of 20 degrees Celsius was set based on a temperate climate and medium altitude, where this is the annual atmospheric temperature. It cannot be used when referring to a tropical or polar climate.

In relationship with the temperature of the human body, mineral water can be classified as:

- hypothermal water, varying between 20°C and 34°C;
- isothermal water, varying between 34°C and 37°C;
- hyperthermal water, with a temperature above 38°C.

According to its osmotic concentration:

Based on the osmotic concentration, the freezing point of mineral water is the same as the freezing point of blood serum (-0.56°C).

According to its freezing point (set by the salt and ion concentration), mineral water can be:

- oligomineral, with less than 1g of minerals per liter;
- hypotonic, with quantities of salt varying between 1g and 8g per liter;
- isotonic, between 8g and 10g of salt per liter and having a freezing point similar to that of blood serum;
- hypertonic, with more than 10g per liter.

This last classification has a great practical importance, as it explains certain effects of mineral water and indicates the ways it should be used in therapy.

Generally speaking, water with less than 15g of salt per liter can be used internally (drinking, enemas, colon cleansing); if it has more than 15g/liter, it cannot be used internally as it irritates the walls of the digestive tract.

The chemical classification takes into account the most common anions and cations, that, when combined, form the main types of mineral water.

Main types of mineral water:

- oligomineral water, with less than 1g of dry residue per liter;

- carbonated water, with at least 1g of CO_2 per liter;

- alkaline water, with more than 1g of dry residue per liter, the most predominant ones being sodium bicarbonate or potassium bicarbonate;

- alkaline-earth water, with more than 1g of dissolved dry residue, the most common ones being calcium, magnesium and bicarbonates;

- ferruginous water, with at least 10mg of iron per liter;

- arsenic water, containing at least 1mg of arsenic salts per liter;

- chlorosodic water, with over 1g of dry residue per thousand, the most common ones being sodium and chlorine;

- iodine water, with at least 1mg of iodine per liter;

- sulphurous water, with at least 1mg of sulphur per liter;

- sulfide water, with more than 1g of dry residue per liter, the most common ones being anion sulfide and sodium and magnesium cations;

- radon water, containing radon.

Drinkable water

On the Earth's surface, water is in a cycle of continuous transformation, but deep down underground there are large reserves of drinkable water.

There are many kinds of drinkable water, but not all of them are good for us. Sometimes it can take a lifetime to know that you've been drinking the wrong water when you could have drank a better one, one that's more useful to your body.

Any drinkable water is good on a short term, but in the long run one must choose the water that has the ideal ingredients for "a recipe of late healthiness", meaning that it keeps you healthy even at an old age.

Usually, the water passes from the atmosphere through the earth before we drink it.

This 'meteoric' water incorporates organic matter or other dry residue that has been dissolved or partly dissolved into it.

After they permeate the earth, they are contained by impermeable rock into underground reservoirs, rivers or springs that flow in various directions.

The earth changes the atmospheric water that enters it, filtering and enriching it with many chemicals from the ground. Water dissolves these chemicals with the help of carbonic acid, forming bicarbonates. The dissolved chemicals may also happen to react between them. It all depends on the type of soil. On the surface, water gets polluted with all kinds of things (rotting organic matter, detergents, sewer water, animal droppings etc.) in areas with human activity. When this water permeates the ground, it acts like a filter and these contaminants remain on the upper levels.

At the same time, the ground oxidizes organic matter with the help of the oxygen within its pores. Organic matter is converted through slow combustion into more basic substances. Carbon is transformed into carbonic acid, hydrogen into water, sulphur becomes sulfuric acid and nitrogen becomes nitric acid, all finally becoming salts that are dissolved by water. After this whole process, the water is now filled with organic substances through mineralization.

Many samples have been taken and research has concluded that nearly all city wells have impurities, whereas underground waters from areas without human activity do not.

Therefore, the quality of surface water (rivers, lakes) highly depends on the lands it flows through. Rivers that flow through uninhabited, rocky regions are cleaner than those flowing in areas with crops or cities, which dump all sorts of chemicals into them.

Factors like oxygen, microorganisms or algae give running water the ability to purify itself. The bigger the volume of water and the faster it flows, the cleaner it becomes.

To make it drinkable, surface water is filtered in water treatment plants. Without going into the details of these filtering processes, we can easily say it's best to avoid constantly drinking tap water. Spring water is normally pure, unfiltered artificially and its molecules are left-centered, whereas the polluted and artificially filtered tap water has more than half of its molecules right-centered. Studies have shown that left-centered water molecules are beneficial, whereas the right-centered ones are harmful, causing premature aging, loss of vitality, slowing down cell activity and more.

The perfect drinking water must be perfectly clear, scentless, with a fresh, pleasant taste and nothing else. Its temperature must be lower than 15°C (59 F) regardless of the season.

Water analysis starts by determining its organoleptic properties. It begins with a test for taste, then for clarity. The scent is more easily established if the water is heated up to 40°C or 50°C (104-114F).

Water hardness can be determined by how fast it produces soap foam. Hard water takes a long time to produce foam.

If you want to find out the water temperature of a well or that of a spring, collect a sample of water in a large recipient (it has to be large so that it doesn't alter the temperature by much) and place the thermometer within.

What's the best drinking water? Obviously, water from a trusted source, like a spring or a well.

A well can be formed of one spring, or it can be the

meeting point of several. The spring can be a seething spring, when water slowly comes out of the ground through sand or dirt particles, or it can be located further underground and found by excavation.

The excavation can be performed manually or automatically (usually for greater depths). Springs can be found at shallow depths (20-25m or less) by digging in wet areas teaming with plant life.

Studying the land can better determine the location of the springs. They're usually found at the base of a hill or when it has a change of slope. The well receives water from upstream; therefore it's necessary that the hill is clean, free of any chemical or microbiological contaminants. It's recommendable that the well is built on land that doesn't have any crops on it, as fertilizer can be a contaminant.

If the land is flat, located in the plains, the drilling is usually done until reaching the groundwater, as any water reserve found higher than that could be only temporary. Groundwater is located at about the same depth of the nearest river or creek. If said river is polluted, the drilling must be done as far away as possible from that area (many hundreds of meters away).

The well can be built with stone or concrete walls. Hardwood (usually oak) can also be used, but the tannin will change the color of the water after a while. It is also recommended that the fountain has a roof.

You can also seek water using the Girari techniques, the Girari being Moroccan water seekers.

Although there is no scientific evidence for this technique, these water seekers search for water using a twig (usually made out of hazel) or a copper ball tied by a thread to a stick.

To avoid digging randomly for water sources, as building a fountain required a lot of effort, the earliest tool man used to find water was a twig. While holding the twig facing forward, parallel to the ground, one walks around on the grounds he wants to dig a fountain on. When the twig begins to slightly vibrate, that means the spring is nearby.

Obviously, the person seeking water must know exactly what he's doing and what kind of source he seeks, adjusting his expectations. Of course, some people get better results than others. They are the people that ask the right questions and act more rationally. Although it doesn't really matter what kind of wood the twig is made from, it's usually made of hazel, probably because the higher density allows vibrations to propagate better. For those of you who want to use this simple tool, it's best to use it with the hand you are most dexterous with. When holding the twig, the contact surface must be as large as possible and the contact pressure as low as possible. The twig must be about 30 to 50 cm long and 7 to 8 mm wide (about the width of a pencil). If it is longer than that, the hands will get weary and disturbances may occur. Beginners are better off with a twig made from hazel that's as straight as possible.

Data has shown that these hazel rods – called radiesthesic tools – were for hundreds of years. A right-sized twig was chosen, of about the same sizes as those mentioned above, and cut on the longest day of the year, while the sun was up.

Immediately after being cut, the twig was debarked and immerged for a few days in a cauldron of boiled herbs. The contents of the cauldron were secret and varied according to the local traditions. Once it dried up, the twig was ready to use.

These rituals were probably a way for the elders to have monopoly rights over this practice, which came with its privileges: social status and high revenue. In the endless attempts to improve this tool, some used a Y-shaped twig and held two ends of it in their hands while facing the third forward. When they felt a vibration, a slight attraction to the ground or other unusual sensations, they saw this as a sign of water. The same desire for improvement led to the invention of the pendulum consisting of a weight hung by a thread. Whether it moved from side to side or in circles, its movements were interpreted accordingly.

Another type of pendulum is the inversed pendulum, made from a ball-shaped weight placed at one end of a very elastic rod. It seems this kind of pendulum is less likely to become affected by involuntary movements of the operator.

Before searching for water sources, we must first assess if the water is drinkable (using the terrain) and establish how deep we want to dig (usually less than 10 or 20m). After that we will scan the field, walking in strips to cover its entire length, to see if we can find a better, more convenient water source.

Depending on the skill of the water seeker (or the radiesthesist), other means of probing can be found. After finding a good water source, we proceed to marking it and move past it, to see how many of them there are. Only after finding all of them we can choose the one that best suits our needs. While passing over the water sources, the radiesthesic tool will visibly move. The spot it moves in will be marked and we will proceed by searching for the spot in which the instrument vibrates the most. We'll mark this spot as well. The greater the distance between the two markers, the greater the depth of the water source. In the case of an underground stream, we will be able to establish its course, flowing speed and any other parameters.

During World War I, Sapper Kelly - an Australian radiesthesist and part of the British expeditionary forces – was in the Dardanelle straits, where drinking water was very hard to come by. Sapper Kelly, used to practicing radiesthesis in his homeland, informed his commanders about his skill and asked for (and got) permission to go and find drinking water. Using two L-shaped copper rods, he quickly located a stream that prevented the British forces from dying of thirst. Sometime later, while being on another mission, he managed to find 32 water sources in a single week.

Be informed!

While on the subject of water, we'll talk about a common label inscription found on water bottles: "microbiologically pure", an inscription that could easily confuse the buyer into thinking that the product contains no microorganisms whatsoever. A well-informed customer knows that obtaining a water that pure requires very advanced technology and that it's more likely that the inscription only refers to germs and pathogens.

The following infectious diseases are associated to drinking water:

- poliomyelitis, a very contagious disease that can lead to paralysis or death;
- hepatitis A, a virus that causes fever, abdominal pain, etc.;
- SARS (severe acute respiratory syndrome), causing muscle pains, intestinal pains, cough and fever;
- Enterobacter infections are the most common, their main effects being typhoid fever and dysentery;
- Botulism, causing breathing difficulties, sight impairment, dry mouth etc.

Water from refreshment drinks

Fruit juices can basically provide a large percentage of the recommended daily water intake.

Fruit juice is defined as being a drink produced by

the squeezing or crushing of fresh, ripe fruits without subjecting it to fermentation.

Fruit juices contain large quantities of water and the fruits they're made from are known to have important amounts of carbs (easily assimilated), acids, coloring substances, flavor substances, minerals, vitamins etc.

The production process has been designed in a way that allows the juice to keep most of the valuable minerals and substances found in the fruit. Heat processing is not recommended, as it turns the molecules from being left-centered to being right-centered. The chemicals in minerals can suffer transmutations inside the body.

Chemicals are known to have double optical activity, meaning they're composed of two substances that rotate the plane of light polarization to the left and, respectively, to the right. It's possible that juice and mineral water have a high level of left-centered chemicals, as the body assimilates them easier. Water that comes from plants has beneficial effects if consumed regularly for long periods of time, as it increases life expectancy. Also, it is good to know that water that comes from fruits or plants has very low levels of deuterium.

Each fruit juice has different beneficial effects on the body:

- cranberry juice is used to treat sight-related illnesses;

- gooseberry juice treats stress and anemia;

- black currant juice is rich in vitamin C and it's recommended in treating rheumatism, arthritis, pneumonia, colds and other conditions caused by a lack of vitamin C;

- red currant juice is recommended to treat liver and renal affections;

- strawberry juice has blood purifying properties;

- apple juice calms the nervous system, enhances the metabolism, grants greater appetite and prevents sclerosis in the arteries;

- prune juice is recommended to people that have a slow digestion;

- grape juice treats anemia;

- hot sour cherry juice makes an excellent remedy for cold;

Generally speaking, drinking juice quenches thirst as well as drinking water does.

Tea (both hot and cold) can be used to replace water in treating certain illnesses. Depending on the plants the tea is made from, it can be used to treat almost all known conditions.

Processed water therapy

In the autumn of 1939, France was at war. Among those leaving Paris, there was an engineer of Arts and Métiers named Achilles Marcel Violet (1886-1973). Having already reached the age of 53, he was excused from military service.

He settled down in Boissy-la-Riviere, in the region of Etampes.

He never lived in the countryside before and his spirit of observation, fine-tuned from all the engineering, made him notice that, after electrical storms, vegetation had an increased rate of growth. He found it curious. He was later informed that the farm animals were somehow more attracted of this water than they were of the tap water they were provided with. When talking this through with his neighbors, experienced farmers all, someone suggested that this may be because storm water is more "active" because of its higher temperature. After doing some research on the matter, this claim was proved false. Still, he thought there might be a relationship between the sudden increase in growth and the water from the electrical storm. He assumed it was due to some chemical compounds that fertilize the soil, but test results proved that false as well. As the storms went and passed, they visibly had the same effects on the crops and animals.

Together with experts from the technical department of the French military, Marcel Violet was studying the way microbes react to different colors of the light spectrum since 1939 and noted that unexplainable changer occur in these experiments. Therefore, the only explanation he deemed possible for the phenomenon was the water's electrical charge.

He discovered that a radiation different from visible light affects these elements.

He repeated the tests and found that water subjected to this radiation seemed to store it somehow. He got the same results regardless if the crop was directly exposed to the radiation or if the water was exposed first and then used to water them. To get a final result, he used cold spiraling tubes to condense the hot fumes obtained from burning hydrogen to obtain chemically pure water. He then transferred this water to a recipient and placed a tadpole within. The agile little creature soon ended up dead. Marcel Violet then stirred the water to oxygenate it and placed a second tadpole in it, which quickly died too. Then he placed the water in a glass bulb, sealed it by melting the glass and placed it on his balcony. After one month, he opened the bulb and repeated the experiment. This time though, not only did the tadpole live, but it also became livelier. The deadly water could suddenly support life. What happened? Chemically, the water had the same structure.

Therefore, the engineer concluded that, doubtlessly, water has an essential role in biology: absorb and then transfer some of these active radiations directly onto living organisms.

Are these radiations vital energy? These experiments should have revolutionized both medicine and biology. The data collected in this series of experiments shows that:

- water has a fundamental relationship with all living things, given its ability to absorb and store life energy;

- energy is a source of life;

- a molecule of water doesn't really support life, but "something else" inside it does: life energy and information.

As we've mentioned before, these experiments should have revolutionized medicine and biology. If they were continued, they would have eventually gathered enough data to paint a complete picture of the human being (a picture which has been only a sketch for millennia).

But let's return to Marcel Violet's experiments.

In his first experiments, Marcel Violet aimed to capture cosmic radiation with an antenna, to separate them special filters and to try and transfer them to certain bodies of water in an effort to come by results similar to those obtained when the water was directly exposed to the radiation. To do this, he used classic dielectric capacitors (artificial) with disappointing results. Then he replaced the artificial capacitors with bees wax. This changed everything. Marcel Violet obtained water which had controllable and measurable accelerating effects on germination.

The bees wax capacitor (12,5 cm long, 12,5 cm wide and 9 cm high) was getting its power directly, through an outlet.

The capacitor was interlaced between the water and the electricity, having one or more electrodes immersed in the lightly mineralized water.

The electrode/electrodes lose some of its/their mass and stain the water.

Water "processed" by Marcel Violet had the following effects on the lab animals:

- they had greater lifespans;
- they had increased viral resistance;

- lab rats chose to drink processed water instead of drinking tap water.

These amazing results applied to plant life as well as animals.

Marcel Violet did the first human testing on himself following a heart attack he had in 1942 (at age 56), drinking one liter each day.

According to him, he did an ECG and found out that the aorta was restored to its normal condition during the four months of 'treatment'.

In time he also noticed the anti-aging effects it had.

From the accounts of Simonne Brousse, she had a beautiful geranium, a plant she cared for very much. It had only five leaves, some of which were so yellow they seemed white.

The next day after being watered with the "electro-vibrant" water, a new leaf appeared. After a month, the plant had seven new leaves and some of the yellow leaves even got back their greenish tint.

The woman found it amazing for a leaf to regain its color instead of withering and fall off.

Many other plants have been experimented on using a device similar to the one described above, a device the

Marcel Violet laboratories in Paris named "The Marcel Violet Bio-oscillator", a high-priced tool they sold for years. From its description, this device incorporated an oscillator which could change the frequency of electrical current to anywhere between 350 and 3500 Hz.

Healers and water

Professor William Tiller presented the following finding:

"By examining the water that has been treated by healers, both spectroscopycally and in infrared, it has been found that the vibration of the water changes, meaning that hydrogen bonds don't link all the water molecules."

The effect was described as similar to the one obtained by immersing magnets in water.

While working with the healer named Olga Worrall on a series of fascinating cases, Bob Miller notes the effects Olga's healing radiations has on a bowl of water.

He tested the water's surface tension and observed a 20% reduction. After a couple of days, the tension turned back to normal, indicating how a measurable physical effect was produced by a non-physical action.

Silvia Demeter declared that water has a very special place in the field of bioenergetical accumulators, being the vital element that's common to all living beings. Therefore, water always has a marginal position in the molecules that form the tissues, ensuring electrical conductivity and energy transfer at a cellular level.

The famous author also says that water incorporates the two elements that have an antagonistic effect on the energetic activity of living matter, oxygen and hydrogen. Therefore, oxygen inhibits charge transfers and hydrogen supports them. One could say that oxygen and hydrogen are a symbol of duality in the living world.

In 1909, Doctor Arthur Lovell was showcasing the discovery that what he named "nerve energy" dissolved into water. As for holy water, we think it's really not necessary to demonstrate their obvious superior quality in relation to usual water.

Magnetized water

Since 1965, Dardimov proved that sunflower, corn and soy grow faster when watered with magnetized water.

Not only that, but they present thicker strains and grow taller than those who are given normal water.

When placing water under a magnetic field, it becomes magnetized. The same effect can be obtained when water is kept in a non-metal recipient for longer periods of time due to the Earth's magnetic field.

Professor Dr. A. Piccardi of the University of Florence claims that magnetized water has a different structure from normal water. Because the water molecule is bipolar, it is enhanced by the magnetic field, forming better quality molecular bonds.

Dr. Bansal of New Delhi explains the way water becomes magnetized: "it's enough that the non-metal bottle is placed on the South or on the North pole of a magnet for the liquid to gain magnetic properties". In reality, the water must be placed between the two poles and the time this process takes depends on the amount of water in the bottle, with a minimum of 15 minutes. Under these conditions, the physical properties of the water change, but its chemical structure always stays the same.

The magnetic effect lasts for about 5 days, but after the first 3 it becomes weaker.

When the magnetic property gets transferred to the living being, it acts on its entire body. Tissues are colloidal solutions and weak magnetic fields are beneficial to biological processes.

Because of its bipolar properties, water guides the body's energetic flow, fixes broken energetic routes, increases the bio-energetic potential and brings balance to cellular energy levels. Homeopaths claim that this water doesn't just cure one disease in particular, but heals the whole body.

Consuming 2-3 liters of magnetized water each days dissolves kidney stones and bladder stones in about 8 to 10 weeks.

Medicinal tea that has been magnetized around 10 to 15 minutes can prove to be a powerful sleeping drug in cases of insomnia.

Another way of treatment with magnetized water consists of drinking only water that has spent 8-12 hours around magnets for two weeks. Then, we pause for about 4 days and repeat the process.

Magnetized water is recommended for all functional troubles and it should be included in the diet for 20 or 30 days, once in spring and once in autumn.

Although there are no restrictions, pregnant women, children or people wearing pacemakers or any electronic implant shouldn't drink it.

It is known to improve the general health condition, as it adjusts the vital functions. The best results were given when magnetized water was included in the diet to treat conditions caused by emotional problems. It can be used to treat conditions like:
- hypertension, cardiac arrhythmia,
- ulcer, colitis,
- dyspepsia, constipation,
- diabetes
- cellulitis and obesity,
- psychological disorders,
- circulation problems,
- breathing disorders,
- sensory disorders,
- metabolic disorders,
- glandular disorders.

In the 18th century, Albert the Great claimed that magnets, as well as magnetized water, eliminates toxins from the body and cures insanity (although there are still some left...). His predecessors – Galen in the 3rd century and Marcel of Bordeaux in the 5th century – recommended using magnets or magnetized water to cure headaches and hydropsy. Paracelsus (14th century), French abbot Lenoble (in 1754), Italian doctor Maiorani and French doctor Charcot (11th century) and many others shared his views. These views are also shared by present-day Asian companies that claim to sell magnets that can cure many illnesses, without informing the public that magnets can also be bad for health.

Let's mention some properties of the water molecules. In liquid state, there are normally chains of 6 molecules, but they can go up to 18 in the rarest cases. The number of links is determined by the polarity. It has been discovered lately that water can exist in a fourth state called liquid crystal, between 0 and 60 C. While in this state, water has a special magnetic charge, the basis for molecular communication. Hydrogen bonds are formed between molecules with a powerful electronegative element that has a small volume and wandering electrons. At freezing temperature, a second hydrogen bond is formed by the oxygen atom, this being the reason for which the ice has a layered structure that determines an increase in volume and a decrease in density. When freezing, water increases in volume by 9%.

In 1781, British physician H. Cavendish discovered that water is formed when a mix of hydrogen and oxygen explode with the help of an electric spark. In 1783, Lavoisier repeated the experiment and, for the first time, accomplished water synthesis. It was then established that 2g of hydrogen combine with 16g of oxygen and form 18g of water. In 1805, Humboldt and Gay-Lussac discovered that water is composed of two atoms of hydrogen and one atom of oxygen.

Ordinary water is formed when oxygen isotopes 16, 17 and 18 combine with the three hydrogen isotopes H, H2 and H3 (hydrogen, deuterium and tritium). This mixture generates 18 different types of water molecules. In its natural state, pure water is a mixture of light water (H_2O), small amounts of hard water (D_2O) and extremely small amounts of very hard water (T_3O). Even in small amounts, their concentrations can change certain properties of the water. Let's see what happens to water exposed to a magnetic field.

According to the author, when exposed to a magnetic field, water becomes polarized and, therefore, increases the solubility for certain chemicals. Superficial water tension decreases. When flowing in underground torrents, water passes many magnetic rocks which transfer some magnetism into it. Water that stays in a fountain, in a bottle or in a glass is magnetized by the Earth's magnetic field. However, water that flows through metal pipes right into our homes loses its magnetic properties.

Solarized water

Solarized water is an easy way to revitalize your body and improve your health. It is obtained by exposing water to the sun for a period of at least 4 hours in a recipient of colored glass, which makes it easier for water to charge with the vibration of that color. Spring water is usually recommended for this process. A glass of this water is to be consumed on an empty stomach or before eating, in small sips. We must not forget that life is light and color and that drinking plenty of water is a problem solving technique, especially if the water has been solarized in a blue bottle. Solarizing with the color blue is recommended for treating various inflammatory conditions. If solarized in a red bottle, the water is said to treat depression and degenerative diseases. If green is used, the water is recommended for hypertension, insomnia, general fatigue etc. If you're having doubts, pour some water into several glasses, cover them and lay them in the sun for 4 hours. You'll see that each glass of water tastes differently, certifying that sunlight has changed the water. Hypertensive or elderly people will start out slow, then progressively consuming up to 1,5 or 2 liters of solarized water. This therapy can be applied for 3 months/year, or it can last a lifetime. It's also recommended that you eat only after one or two hours after drinking the curative water.

Living water therapy

We know that when a chemically neutral liquid (like water) is electrolyzed, there will be an acid environment forming around the anode and an alkaline environment will form around the cathode. When we interrupt the process, the liquid becomes neutral again, as the movement of the molecules makes the two environments mix together.

Nowadays, an electrolytic bath is equipped with a semi-permeable wall (it can simply be a canvas or a fabric that separates the anode from the cathode). This wall doesn't allow the products of the anode and cathode to mix. Experts from the Institute of Physical Chemistry and from the Institute of Electrochemistry of the former USSR were developing the EDG3 device, a machine that produced an oxidizing substance that had amazing healing properties. When it was tested on superficial wounds, the skin healed quickly. By asking their test subjects to gargle the substance, researchers were able to cure a form of tonsillitis that had resisted previous antibiotic treatment within 24 hours. This substance also slowed tumor growth and stimulated the oxidizing of the liver.

Valeri Serghievo, Doctor of Medicine, who was one of the creators of this device, has pointed out that the human body is equipped with 3 systems that metabolize toxins: the immune system, the liver's oxidizing system and the liver's eliminating system. There was some earlier research that produced artificial oxidants for the first two systems.

Now, an artificial oxidant was finally made for the third system. The new method of treatment was named 'the detoxification of the body through the use of the EDG3 device'.

Living water is obtained through the use of a device made from:

- 2 stainless steel electrodes pinned to an insulating support with 35 mm between them.

- An electrical cord with a plug for easy access to the power outlet. The cord supplies the electrodes with electricity through a bridge rectifier in such a way that one gets its electricity at the positive end and the other at the negative end.

- A recipient made out of glass or plastic.

- A small bag made of impermeable tarpaulin (like the one they use in tents).

Water is placed in the recipient, the bag is tied around the anode (the electrode tied at the positive end) and the electrodes are immerged. The device will remain plugged in for 5-7 minutes and then it will be unplugged. The electrodes are removed, and the water from the bag will be poured into another recipient. The bag is then filled once again with water and the process recommences.

The water in the bag (or the 'dead water') will be yellowish in color, with a pH of 5 and a lightly acid taste.

The water in the jar (the living water) is alkaline. It has beneficial effects on the body, not only curing disease, but also restoring the body to its normal levels. The acid water can be used to kill bacteria, molds, pests etc.

Verified treatments:

- Tonsillitis: gargle acid water and then drink a quarter glass of living water five times a day for three days.
- Ankylosis, the stiffness of the joints. Drink half a glass of acid water before eating six times a day for two days.
- Stomach burns: drink half a glass of living water. The burning sensation disappears and gas is eliminated.
- Diarrhea: drink half a glass of acid water.

- Prostate: drink half a glass of living water 30 minutes before eating four times a day for 3 days.

- Sterilizing: the object is immersed in acid water.

- Coughs: four times a day for two days, drink half a glass of living water after eating.

Living water has a pH greater than 7 and the beneficial results it has on the body could be due to the alkalization of the body, as it's also known as alkaline ionized water.

The magazine Ion Life Detox News explains in its 7th number in 2004 the reasons we get fat, sick and old. In Japan, one family out of five drink ionized water.

It's also believed that acidosis is connected to conditions like:
- cardiovascular injuries,
- obesity, diabetes,
- cancer,
- autoimmune deficiencies,
- premature aging,
- osteoporosis,
- fatigue,
- low levels of energy.

Ionized alkaline water is a powerful antioxidant, supplying the body with oxygen and energy. Another feature is that alkaline water has hydroxyl ions (OH), meaning oxygen molecules that have an extra electron attached, as do all antioxidants (vitamin C, A or E).

When hydroxyl ions react with free radicals, the latter are neutralized and the body is supplied with more oxygen.

Oxygen destroys the viruses and bacteria attacking our bodies and eliminates the acid residues from the tissues. It's considered to be the most important nutrient, as it's vital for our bodies.

Therefore, even if you're healthy, it's recommended that you drink a glass of ionized water every morning before eating.

Distilled water

Distilled water is water obtained through vaporizing and condensation. It doesn't have mineral salts and cannot support life. Some therapists include it in diets, as it has detoxing properties. Drink about 2 liters of distilled water in the first day. The toxins (about 2-5% of them) will be eliminated with the water. It's recommended that you ask your doctor's advice before pursuing a longer diet.

Pi water

Pi water was discovered in 1964 by the Japanese professor Shoi Yamashita while he was studying certain phenomena that occur in botany. He found that small amounts of ferro-ferric salt are responsible for budding. Pi water is produced by immersing this salt in water and then filtering it through a ceramic filter.

The ferro-ferric salt (the bivalent and trivalent ferrite) remains in homeopathic concentrations in the water. The particles are very small, conductive and responsible for information transfer at a cellular level.

Pi water boils faster and freezes somewhere between -1°C and -7°C.

In 1998, at the International Symposium on Cancer Prevention in France, the following conclusions were drawn:

1. Pi water is very similar to the water in our own bodies.
2. It acts as an antioxidant, as it destroys free radicals.
3. Energizes. Fish live longer when placed in a closed tank filled with Pi water.
4. Increases the activity of antibodies.
5. Improves cellular healing.

6. Carry more oxygen throughout the body.

7. Is an excellent detox agent.

8. Pi water consumptions helps slow the aging process.

Doctor Tokafumi Tsurumi used Pi water in treating the following conditions:

- Stomach cancer, shrinking brain tumors, leukemia.
- Dermatitis and alopecia.
- Diabetes.
- Quick recoveries after heart attacks.
- Strengthening the immune system.

Kangen water

The word "Kangen" is the Japanese word for "going back to the origins". Therefore, Kangen water is the water found in nature before all the pollution. It's lightly alkaline, ionized, antioxidant, restructured and rich in active hydrogen.

The molecular formula for Kangen water is $H+OH$. It's considered living water, as it washes away the toxins and neutralizes the acids in the body. It's a powerful antioxidant, maybe even more powerful than vitamins, for the active hydrogen produces large quantities of electrons inside the human body.

Kangen water is made by the Enagic Company, founded in 1974 in Tokyo. The company also produces alkaline water devices for other types of water since 1982.

In the past, people went as far as the Himalayas to drink this alkaline water with a hexagonal structure, very good for preventing diseases.

With this in mind, Enagic set up to produce household appliances so that everybody could produce their own alkaline water.

Enagic has OEM certification (Original Equipment Manufacturer), which means that all parts of the devices they build are manufactured and assembled in Japan exclusively by the company.

The equipment is licensed as medical equipment by the Health Ministry of Japan. There's an ongoing campaign to introduce these products on the European market as well.

Enagic devices produce 5 types of water:

a. 3 types of Kangen water with a pH level between 8,5 and 9,5.

b. Neutral water with a pH of 7,0.

c. Acid water (or beauty water, with a pH level between

4 and 6).

d. Highly acid water (disinfecting water with a 2,5 pH).

e. Strong Kangen water (detergent water with a 11,0 pH).

Silver water

According to some research made in the old USSR, human tissue has 0,02 mg of silver per 100g of dry tissue. Silver is incorporated through food and, probably, through the transmutation of some elements. Lately it has been found that food contains less and less silver. In high concentrations, silver is found in the brain tissue, in the nucleus of the nerve cells etc. Lack of silver in the body can lead to a series of illnesses.

As for how the body uses the silver, it's hard to tell. It's possible that silver increases electrical conductivity in the cells.

Silver can be included into the diet by drinking colloidal water, a type of water that has silver ions under colloidal form. Colloidal particles are the smallest particles matter can be decomposed to without losing its properties. These particles are found in water and are electrically charged. Because the same types of charge repent each other, they are always floating around. Colloids play a key role in nature, for all natural processes are based upon colloidal structures. Because of their small size, colloids can reach and affect even the smallest places, like cells.

The existence of a colloidal system is based upon the following criteria:

1. Have two components (here: water and silver).

2. Have three different phases: liquid/solid, gas/liquid.

3. The substances must not dissolve one another.

Colloidal silver is made out of silver particles that are positively charged in the water.

It represents a dispersion made in water by the process of electrolysis.

To produce colloidal silver you take two silver electrodes and immerse them in water while applying an electric current to them. Once the electricity is running, an electrode can attract silver particles varying in size from 0,001 μm to 0,01 μm and contain between 1 000 and 1 000 000 positively charged silver atoms. Some of them are even visible in the water, being similar to a golden or silvery fog. The concentration of silver ions depends on the distance between the electrodes, their surface, the intensity of the current that passes through them and the time they're subjected to it. Silver particles get positively charged and reject each other in the water in such a way that they keep afloat and spread evenly among it. This spreading movement is random and visible under a microscope.

The silver concentration varies between 3 and 50 ppm (parts per million), meaning 3 to 50 milligrams of acid for each liter of water.

The particles lose their electric charge in time and tend to cluster. It's good to know that colloidal water is best consumed at a short while after it's produced. Many countries have conducted experiments proving that colloidal water has the following qualities:
- bactericidal
- destroys viruses
- antifungal
- antioxidant

These effects are caused by the fact that cellular membrane is negatively charged, attracting the silver ions that block the cell's ability to reproduce and inhibit the enzymes in the breathing organelles, thus killing the bacteria.

Certain American universities discovered that colloidal water has destroyed infamous bacteria like Staphylococcus Aureus and Salmonella.

Research has confirmed the experimental results of Dr. Henry Margraff, a scientist that demonstrated around 100 years ago that silver can destroy over 650 types of pathogens in a matter of minutes. Dr. Henry Krunk said: "I'm not aware of any microbe that cannot be destroyed in less than six minutes using colloidal silver". Viruses, albumin molecules 0,02 to 0,01 µm in size reach the inside of the cell through osmosis, then reproduce and kill the cell.

Silver ions about 0,001 µm in size go through the cellular membrane of the virus and, once inside, block the enzymes that supply it with oxygen. Without it, the virus withers and dies. Free radicals are negatively charged molecules, having one or more additional electrons. Healthy cells have a neutrally charged membrane. As free radicals want to become neutral themselves, they attack the first healthy cell they come across and transfer their additional electrons to it.

The initially healthy cell becomes negatively charged and, therefore, unstable. At its turn, the cell affects another healthy cell, causing a chain reaction that makes all the healthy cells die. These clusters of free radicals are the cause for diseases like cancer, Alzheimer's, cardiovascular diseases, diabetes, autoimmune conditions etc. Antioxidants are chemicals that take the extra electron away from the free radicals, becoming stable themselves and preventing the deadly chain reaction. Colloidal silver contains positive ions that take the electron away from the free radicals.

It has been discovered that colloidal silver stimulates blood production, thus increasing the number of erythrocytes, lymphocytes and hemoglobin.

Colloidal silver is neither toxic, nor allergic. Any surplus is easily eliminated within a day and doesn't clutter up the body. It doesn't interact with any of the drugs normally taken during treatment. Colloidal silver can be produced or bought at concentrations between 5 ppm and 25 ppm. It must be kept away from sunlight and household appliances. It must never be kept in the refrigerator.

If you want to follow a diet of colloidal silver, it's recommended that you consume 3 tablespoons per day. Hold the water for about 20 seconds under your tongue. Don't eat anything for 30 minutes before or after drinking it. When treating certain illnesses it's recommended that you take 1 or 2 tablespoons every 2 hours, or even 1 or 2 spoons. The dosage is as follows:

- Small dose – 1 or 2 tablespoons once a day.
- Medium dose – 1 or 2 spoons taken twice or thrice a day.
- Large dose – 50 to 100 ml a day.

As a measuring aid, a spoon has somewhere between 7 and 10 ml and a tablespoon has about 3 or 4 ml.

There are no known side effects. At the beginning of the treatment, chills or fatigue sensations may appear. Colloidal water must not be mixed with water or anything else.

In cosmetics they use water with colloidal gold, which is produced much like colloidal silver water.

When the American states were newly-formed, the citizens used to put a silver, 1$ coin in a water bowl for sterilizing it. One day someone discovered that the water had an odd taste and had beneficial effects on the body. After 1934, when the first antibiotic was discovered, silver was no longer used in treatment and its price went down a lot. Antibiotics are obviously better than silver water, but they're used only when disease occurs, not to prevent it.

In a familiar manner, water with colloidal gold or other metals has various uses.

Water enriched with coral calcium: CORAL-MINE

Calcium is an essential element in the body's functioning. Coral calcium is obtained from fossilized coral, like the white Sango coral.

Calcium ions are more easily assimilated by the human body.

Shungite water

Shungite is a type of rock originated from a large meteorite that fell tens of thousands of years ago near Onega lake, in the Karelia Republic (NV Russia). Its therapeutic properties are known for many centuries. In the 18th century, tsar Peter the Great commanded his people to build a resort near the spring that passed through this Shungite deposit with the name of Martian Waters. He also commanded that every soldier must always carry a piece of Shungite rock and place it in their water to disinfect it. It's said that Shungite protected the tsar's soldiers from epidemics like dysentery.

In the 20th century, scientists analyzed the

Shungite rock and came up with an explanation for its miracle properties. Shungite purifies water of nitrates and nitrites and remove the excess of elements unfit for human consumption. At the same time, it saturates the water with calcium and magnesium salts to the optimal concentration for the human body. Research has shown that water mixed with Shungite has bactericidal properties after two or three days. Shungite water is a colloidal solution that affects the body on more than one level. It's recommended to prevent and treat conditions like anemia, allergies, asthma, gastritis, liver and kidney disease, gallbladder problems, thyroid issues etc. There are no contraindications for Shungite water. Usually, 200 grams of Shungite are mixed with 2 liters of water and the person consumes 2 or 3 glasses a day. The water shouldn't be stirred and the Shungite rocks are to be replaced anywhere between 6 or 12 months of use.

Treatment with clay that has been dissolved in water

Ancient Egyptian papyruses, Greek and Roman medical manuscripts and the unwritten knowledge of various Asian, African and Native American peoples all mention the curative properties of clay and the way it helps treat various internal swelling, infections, skin burns, abscesses and digestive problems.

A natural, living earth, clay has proven to be really useful in combating dysentery, a disease that was laying waste to the soldiers camping in the trenches in World War I.

Clay may have become such an important part of traditional medicine because humans would have studied the instinctual behavior of animals and noted that, when they have some kind of injury, they fervently search for clay-filled swamps in which they immerse their afflicted parts.

When the affliction is internal, not external, animals were licking limestone or drank water from swamps that had limestone around them.

The acclaimed scientist Raymond Dextreit has some amazing information about this ancient curative agent in his book, "L'argile qui guerit", published in Paris in 1978.

When using clay to treat a festering wound, it heals surprisingly fast, although there is currently no scientific explanation behind it to this day, explains R. Dextreit.

Especially when intended for internal use, bits of clay

are added into the water.

After cleaning it for dust and other debris, the clay is crushed to a fine white, green, yellow, red or grey powder.

The water with crushed clay is prepared a few hours before its intended time of use by dissolving a tablespoon of clay powder into the water and stirring it with a spoon.

This amount represents the recommended daily dosage throughout the 3 weeks of diet. Some therapists recommend that the water should be drunk only once the clay has dissolved completely, other say that the mix should only be shaken for a while and then drank.

The mixture is drunk in the morning, about half an hour before eating.

This mixture helps in the treating of the following conditions:
- hepatic and biliary insufficiency
- gastroduodenitis and ulcer
- nephritic colitis
- depressions and nevrosis
- lack of minerals

Also, water springs located at the base of hills that contain clay have water with about the same properties as water mixed with clay.

Rainwater dissolves the clay found in the hills and carries it to the water pools. A good indicator for water mixed with clay would be the traces of clay at the bottom of the pool.

Clay-filled water usually leaves a greyish tint on the recipients in which it was poured.

Clay-filled spring water also is used to treat kidney stones, as it has about the same therapeutic values as mineral water:

- both waters have either an alkaline pH or an acid one, which inhibit the development of kidney stones if the acid environment of the bladder is favorable to them,

- both waters have about the same osmolarity (a unique concentration of ions or minerals that help dissolve and eliminate the kidney stones),

- clay water and mineral water have certain ions that have a favorable influence on the composition of urine.

Alkaline spring water contains salts, the most common being bicarbonate ion.

Because they're hard water, their hardness increases as the spring gets deeper, as the water has to pass through several layers of clay.

People that live in areas near clay water have no history of kidney problems. That's because the effect this water has on the body can only be seen after prolonged consumption (at least 30 days). Unlike mineral water, that loses some of its properties unless it's drunk directly from the spring, clay water can be bottled and consumed afterwards. Also, there aren't any restrictions as to the amount of clay water we should drink.

Water and deuterium

Deuterium is a hydrogen isotope that has the same behavior, but a larger mass. Obviously, water formed with deuterium instead of hydrogen will be heavier, therefore the name of 'heavy water'.

In a tissue made from ordinary water, heavy water molecules "are sagging" like stones in a bag. Now that couldn't be good, could it?

Hard to say, but some research has shown that heavy water has negative effects on organic tissue.

Normally, water doesn't contain many heavy water molecules and even those are hard to come by, as they're a product of fractional distillation. Heavy water also has different properties: it dissolves salts at a slower rate, reacts slowly than regular water and it's considered a toxin.

However, an exchange of isotopes between the hydrogen in the body and the deuterium can be made. Research has shown that heavy water inhibits the sprouting process.

Experiments on lab rats has shown that a 15% concentration of deuterium in drinking water can slow the normal growth of the fetus, a 25% concentration can change the fetus's central nervous system and a 50% concentration the embryos go through a resorption process.

A 30% concentration has even made the mice sterile. The effects of deuterium depend entirely on the animal's strength and sensibility.

That being said, the following changes usually occur: there's an increase in density, the steam pressure drops and the dissociation of weak acids suffers major changes, as do the bonds of hydrogen. Isotopic substitutions within the natural macromolecules affect their tridimensional structure. If this structure changes, the enzyme activity also changes and, with it, all the reactions that take place in the living cell. These changes could be explained either by a change in the kinetics of biochemical reactions or by slight modifications occurring in the microstructure and the morphology of the cell as a direct effect of heavy water. A series of important research on the field of enzymatic reaction kinetics has shown that enzymes react slowly in heavy water than in regular water.

It has also been shown that humans replace about 7,7% of the total amount of water in the body on a daily basis, but heavy water takes longer to eliminate.

Therefore, if the water source contains heavy water, your body will act as a filter, letting the regular water pass through and keeping the heavy water.

Water low in deuterium

Water with low levels of deuterium (25 ppm) can be obtained through repeated distillation and some countries (Romania, Hungary, Japan) use it in the treatment of cancer. Data gathered from several experiments has shown that, out of 500 chickens that consumed such water, none died and all grew faster and stronger than those who drank only regular water, for it enhances intercellular communication. This water has a similar structure to water that has melted off a glacier.

Properties:
- it's structured in micro and macro clusters
- energizing

- antibacterial
- compared to regular water, it provides 3 to 7 times the normal vitality

It can be homemade with the Japanese device called the Living Water Generator.

It's said that regular water "memorizes" the path it took to the surface, all the rust and 'garbage' along the pipes it goes through, the properties of the earth layers, as well as all the "events" it witnesses.

One can erase this memory through distillation, freezing or with the above-mentioned device.

The most eloquent examples of this water's beneficial effects are the people living in mountain regions. They rarely get sick because they drink water that comes from the melting of snow and glaciers.

That's why people residing in Northern India, Northwestern China, Southern Tajikistan, Northern Pakistan and Eastern Kazakhstan live longer lives.

According to the locals, they live an average of about 90 or 120 years.

Hunza – a miracle

Long before trying to restructure ionized water, several scientists noted that the Hunza region's water and diet are somehow special.

The people of Hunza lived longer than the rest of the world.

Women here were giving birth even at an older age. People didn't have cancer, dental cavities or degenerative diseases and nobody could say why. It was later found that the water of the Hunza region was responsible for this high longevity rate.

This water came from a glacier. As they couldn't recreate the conditions anywhere else, the scientists remained in this region and studied the unique water structure in an effort to recreate it.

They found out that the nature of water was alkaline, with high levels of active hydrogen, negative redox potential and high levels of colloidal minerals.

Although the original story was lost in time, it's believed that Russian and Japanese scientists also probed and researched the water in the Hunza region after World War II.

This research brought a lot of attention to the region

and inspired the romantic novel "Lost horizon", a tale of a mythical kingdom named Shangri-La, a place where people lived forever.

Under Frank Capra, this novel became a successful movie that portrayed the isolated life in the mountainous regions of the Himalayas, where people told each other stories about secret springs.

By 1948, although the tales of immortality weren't proven by anyone, everybody was hoping to understand the reasons people in Hunza lived for such long periods of time.

Then, scientists came to the conclusion the secrets were in the water and the diet.

The Japanese have found new electrolysis techniques based on some experiments the Russians did.

Alkaline ionized water is obtained by separating the water into two components: acid and alkaline.

The Japanese built the first ionizer in the early 1950s.

After a while, Japanese doctors confirmed the curative properties of the water and, in 50 years, it's estimated that over 30 million Japanese have used the devices that produce homemade alkaline water, a substance we've called 'living water' in one of the previous chapters.

Today, some companies produce and sell Hunza water. For example, if you have 20 euros to spare you can buy a 60ml flask that contains mineral concentrate of liquid nanoclusters. The mixture is then poured in the water and improves the plasmatic environment of the cell, its pH, the superficial tension and its conductivity.

The presence of nanoclusters in the water also improves the body's ability to assimilate nutrients. The water's superficial tension drops and it resembles more and more to the cellular liquids inside the body. The liquid is a powerful antioxidant and it neutralizes the negative effects free radicals have on the body during its metabolism. If a person chooses to eat fat and sugar or exposes himself/herself to toxins, polluted air or water the number of free radicals increases.

As for the dosage, it's advisable that 8 drops of mixture be added to a glass of water.

After 30 years of research, Dr. Patrick Flanagan succeeded in synthesizing the specific compound of Hunza water and called it Crystal Energy.

He was also the inventor of the so-called technique to obtain these small particles, the microclusters. The Flanagan microclusters are the most powerful antioxidants yet.

They destroy free radicals and improve the biological environment that surrounds the cell: pH, superficial tension and conductivity.

The discovery of electrons in Hunza water led to the birth of a new product called microhydrin.

So what kind of water should we drink?

Firstly, water that comes from a great depth should be avoided, for it contains large quantities of heavy water. In time, heavy water has 'sunk' and the regular "light" water has seeped its way to the surface. Therefore, surface springs should contain the least amounts of heavy water (and by surface spring we mean a spring that's a few meters deep). Still, there are cases of water that has even less heavy water concentrations.

One example could be the water inside plants or fruits.

We can rest easily knowing that juices made from plants and fruits are healthy (and healthier still if they haven't been chemically enhanced). It's highly unlikely that you'll steer clear of all heavy water molecules for as long as you live, but it's important that you don't ingest too many and too often. Therefore, it's important to choose the water you'll drink in the long term, not in the short term.

Regular water

Who or what brought so much water to our planet? Why? Were there always so large amounts of it? Nobody can tell exactly how it all formed here. It will remain a mystery as great as its properties.

Nothing is as malleable and adaptable as water. Still, water erodes even the toughest structures and nothing can defeat it, although anyone can conquer it.

"That which is flexible and adaptable conquers that which is strong, and that which is soft defeats that which is hard. We all know it, but no one dares to live like this" were the words Lao Tse used to describe water 2500 years ago.

Since man first laid eyes on it, water was always an object of study.

It has physical and chemical properties unlike any other liquids.

No one could explain why water density rises below the freezing point and drops above it, when any other substances contracts when frozen and expands while heated.

Water has a small molecule with feats no other molecules can boast. If even one of those feats weren't there, life on Earth couldn't exist.

Each of its properties is unique and represents an exception to the rule (in this case, the generally accepted laws of physics).

Science can't tell us why water is the only substance on the planet that exists in three separate states; why it has the biggest superficial tension of all liquids; why it's the most powerful solvent on Earth and, last but not least, why it's the most common liquid.

Isn't it odd that water can rise through the trunks of massive trees, defying gravity and withstanding tens of atmospheres of pressure?

In the 1980, someone came up with a fantastic theory: water has a memory. Several countries have conducted research and showed that water analyses and records every external influence and then "remembers" all that information about the surrounding environment. Therefore, any substance that comes in contact with water leaves its trace in its structure.

Our ancestors have anticipated this property well by using silver bowls to turn regular water into sterilizing water.

Today, this water is still used as an antibiotic in Afghanistan and Iraq. The US army uses water that has

1 silver atom per 100 molecules of water to treat wounds and destroy any germs that lie within.

US presidents used such a water to disinfect their hands.

So what's it like? As it receives new information, water also receives new properties, all while maintaining the same chemical composition. An old theory was arguing about the importance of the water's chemical composition. However, it may be that it was stupid to believe that, because water's taste changes although the chemical formula doesn't.

Its structure is way more important than its composition.

The structure is given entirely by the arrangement of the molecules. The water molecules usually come together and form molecular groups called clusters.

Scientists have theorized that these clusters work like some kind of memory cell, being the recording mechanism of water. Of course, water is still water. It's only its structure that reacts to change, similar to a nervous system.

If we think of a cluster as only a group of molecules, that would mean it can only live short periods of time. However, if we consider it to be a structure from which molecules come and go, the cluster can support itself for very long periods of time.

The stability of cluster-type structures suggests that water is, indeed, capable to record and store information. The only thing that's unsettling is that we don't know exactly how it does the information transfer.

The information memory of water is similar to the process of locating letters in the alphabet.

Water can also sense human presence and human emotions, those being the most powerful influence humans can have on it.

Professor Korotkov has done many experiments to show the effects human emotions have on water.

With the help of modern technology, water can be given a structure artificially.

Using structured water makes vegetables grow faster and reduces the water necessary for irrigation by 20%.

It's widely-known that big cities have a closed-circuit water system.

After being physically and chemically purified and filtered, the water is recycled as domestic water.

The most worrying type of water pollution is informational pollution, as water gathers many information as it passes through miles and miles of pipes.

We don't pollute only chemically, but also structurally,

at a large scale. Water is nearly dead before it enters our bodies. It's true! If you don't believe it, compare it to the water that you can drink directly from the spring, like our ancestors did.

At a molecular level, water creates the structure of the DNA. The DNA spiral wouldn't exist if it weren't for water. It's also a key component in the structure of protein. Every seed, every embryo, they all start their life in water. Water is the operating system of a computer called "LIFE".

In the dialect the Peman tribe of Venezuela uses, "Roraima" means "the mother of all water". A group of biophysicists have travelled to this destination in 2005 to collect a sample of water that they say man came in no contact with.

Such water only exists in that place.

Some theorize that a continent named Gondwana existed in the Southern hemisphere during the Paleozoic Era.

3,5 million years ago, powerful tectonic movements split the continent into several pieces, thus forming the high plateaus the Peman call 'Tepuis', meaning 'pillar'. Roraima is the biggest, a place that's almost impossible to reach.

Above this plateau, there's always a big cloud. As

night falls and the temperature drops, it transforms into a light fog and, when the moon rises, it gains a blue glow.

While the fog is glowing, one can see the drops of water suspended in mid-air.

At the slightest breeze, these water vapors transform into raindrops and form the water streams that cascades down into the valleys.

Research has shown that Venezuelan water is about 40 000 times more active and quickly energizes the human body.

The natives live long lives, are happy, full of energy and don't want civilization to reach them. Here's an older, different case:

Autumn 1632, a farmer named Ganz from Enengen, in the Hessen district, started his journey to Southern Italy to seek a better grazing spot.

His journey passed through Walschandt am Rhein, belonging to the parish of Konstanz.

Suddenly, Ganz had a feeling that this place was somehow connected to him and saw water springing from the ground. He bent over and drank the spring water. Many years later, he told his nephews how the water restored his memory and reminded him that he was in the place he was born. Modern science claims that the water structure in the mother's body is identical with the water that goes to the infant in his first days. Therefore, there's a bond between us and the place we're born in, a bond that stays with us forever, a bond that depends on water.

Water is different wherever you go

As it rushes through the surface, passing rocks and minerals on its way, water assimilates all the information they carry.

Bottled water is usually 'dead' water, although it tastes good and healthy. People cannot see the difference between naturally pure water and an artificially purified one.

However, an animal will always pick the spring water if given a choice.

Recent discoveries seem to suggest that spring water has a new property: it can catch fire.

The water itself burns, as it has a special structure. It burns slowly and it's only detectable using special gear.

Personally, I think that it may contain some dissolve gasses which could ignite. These gasses could be released after a while, thus stopping the burning process.

On the other hand, it's a known fact that something (anything) burns only if it contains a tiny amount of water.

In closing, if some of the things you've learned about

water seem downright crazy, remember an old Persian legend that goes something like this:

There was once a wise man that said a day would come when all the water in the world will disappear, with the exception of that in water reserves, and be replaced with another type of water that will make anyone that drinks it go insane.

Only one man believed the prophecy and started stockpiling water. When the foretold day came, all the water in the world had disappeared. The man that listened to the wise man drank only water from his reserve, while the rest of the world drank the new water. After a while, they all went mad and the man that had water reserves was the only sane one. Everyone else started calling him a madman, for he wasn't like them.

Once he became aware of that, the man succumbed and drank the new water like everyone else did, becoming mad himself.

Everyone was glad he was a healthy man like them, although they were all mad now.

So I ask you, dear reader, do you know what kind of water you're drinking?

Bibliography

1. Alexan Mircea, Bojor Ovidiu, Fructe şi legume – factori de terapie naturala, Editura Ceres, 1992.

2. Alexander Hellemans, Bryan Bunch, Istoria descoperirilor ştiinţifice, Editura Orizonturi, Bucharest

3. Alexandru Roşu, Terra, geosistemul vieţii, Editura Ştiinţifica şi Enciclopedica, Bucharest, 1987.

4. Andrei Rădulescu, Electro Terapie, Editura medicala, Bucharest, 1993.

5. Beral E., Zapan M., Chimie anorganică, Editura Tehnică, Bucharest, 1997.

6. Brahmachary Icvaracharia, Magnetism hindus, Editura Mirador, 1992.

7. Chatterji J. C., Filozofia ezoterica a Indiei, Editura Princeps, 1993.

8. Demetrescu Scarlat, Din tainele vieţii şi ale Universului, Editura Emet, 1993.

9. Dramba Ovidiu, Istoria Culturii şi Civilizaţiilor, Editura Ştiinţifica, Bucureşti, 1990.

10. Dumitrescu Raul, Taina Longevităţii, Editura Miracol, 1996

11. Egelstaff D. A., An Introduction to the Liquid

State, Academic Press, London, New York, 1967.

12. Fayman R., Fizica moderna, Editura Tehnica, Bucharest, 1969.

13. Florescu M., Enigmele şi paradigmele materiei, Editura Politica, Bucharest, 1993.

14. Fritz Elsner, Die Praxis der Chemikers, Hamburg, 1895.

15. Georgescu L., Fizica, vol. I, partea VI, Editura didactica şi pedagogica, Bucharest 1972.

16. Girard Ch., et Dupre A., Analyses des Matieres Alimentaires et recherche de leurs falsification, Paris, 1894

17. Green H. S., Molecular Theory of Fluids, North Holland, Amsterdam, 1952.

18. Hirschefelder J. O., Curtis Ch. & R. B. Bird, Molecular Theory of Gases and Liquids, John Wiley, Nework, 1954.

19. Holodov I, A., Magnetismul în biologie, Editura Ştiinţifică, Bucharest, 1974.

20. Hristenco D., Radiestezia, Editura Teora, Bucharest, 1995.

21. Ivan Sabin, Terapii naturale de vacanţă, Ghid terapeutic naturist, Editura Sport, Turism, 1969.

22. Kikoine, A. Kikoine I., Physique Moleculaire, Edition Mir, 1979.

23. Mamularu Gh., Cura hidrominerala în litiaza urinara, Editura Medicala, Bucharest, 1986.

24. Mihaiţă Toma, Apa, miracolul vieţii, Editura Dharana, Bucharest, 2010.

25. Percek Arcadie, Terapeutica naturistă, Editura Ceres, 1987.

26. Pinnkep E. V., Podzemnaia ghidrosfera, Izd „Nauka", Sibirskoe otdelenie, 1984.

27. Pisota I., Buta I., Hidrologie, Editura Didactica şi Pedagogica, Bucharest,1983.

28. Philipe Aries, Georges Duby, Histoire de la vie privee, Editions du Seuil, Paris, 1987.

29. Poltzer A., Chimia şi analiza alimentelor şi băuturilor, Librăria SOCEC & Co, Bucharest, 1899.

30. Popescu Octavian, Valeriu Popa, Mit sau adevăr, Edimpex Speranţa, 1990.

31. Richard Lewinsohn, Histoire des animaux, Librairie Plon, Paris, 1953.

32. Rottger H., Kurzes Lehrbuch der Nahrumghsmittel – Chemie, Leipizig, 1894.

33. Sahleanu V., Chimia, fizica şi matematica vieţii, Editura Ştiinţifică, Bucureşti, 1965.

34. Spikes J. D., Photobiology, Acad. Press., N, Y.,

1968.

35. Stefan Airinei, Pământul ca planetă, Editura Albatros, Bucharest, 1982.

35. Tarasov L. V., Laserii realitate şi speranţe, Editura Tehnică, Bucharest, 1990.

36. Th. Weyl, Handbuch der Hygiene, Jena, 1897.

37. Ujvari I., Current concepts about establishing parameters for physical – geographical factors of hydrological processes, Beitrage zur hydrologic.

38 * * * Enciclopedia Universală Britanica, vol. 1, Bucharest, 2010.

www.ingramcontent.com/pod-product-compliance
Lightning Source LLC
Chambersburg PA
CBHW061514180526
45171CB00001B/179